# Level 3

# Problem Solving

Donna Marie Mazzola

Bright Ideas Press, LLC
Cleveland, OH

# Summer Solutions Level 3
Problem Solving

All rights reserved. No part of this publication may be reproduced or transmitted in any form or by any means, electronic or mechanical, including photocopy, recording, or any information storage or retrieval system. Reproduction of these materials for an entire class, school, or district is prohibited.

**Printed in the United States of America**

ISBN-13: 978-1-934210-53-6
ISBN-10: 1-934210-53-6

*Cover Design*: Dan Mazzola
*Editor*: Kimberly A. Dambrogio
*Illustrator*: Christopher Backs

Copyright © 2010 by Bright Ideas Press, LLC
Cleveland, Ohio

# Summer Solutions Level 3
## Problem Solving

| Lesson | Topic | Page |
|---|---|---|
| 1 | Strategy: Make an Organized List | 2 |
| 3 | Strategy: Guess and Check | 6 |
| 5 | Strategy: Look for a Pattern | 10 |
| 7 | Strategy: Draw a Picture | 14 |
| 9 | Strategy: Work Backward | 18 |
| 11 | Strategy: Solve a Simpler Problem | 22 |
| 13 | Strategy: Make a Table | 26 |
| 15 | Strategy: Write a Number Sentence | 30 |
| 17 | Strategy: Use Logical Reasoning | 34 |
| 19 | Skill: Combining Strategies | 38 |
| 21 | Skill: Sort Angles-Acute, Right, & Obtuse | 42 |
| 22 | Skill: Use a Graph/Make a Graph | 44 |
| 24 | Skill: Notice Unimportant Information | 48 |
|  | Help Pages | 63 |
|  | Answers to Lessons | 77 |

Dear Girls and Boys,

I am Math-U and this is my friend, Add-E. We're going to be around every summer to help you with Problem Solving. Math problems with words are tricky sometimes. You have to come up with a plan or strategy for solving them. You learned lots of strategies last year and we don't want you to forget them. So, this summer we will review nine strategies. I'll be showing up on the pages and even in your work boxes to give you hints. Look at what I'm doing in the cartoon. Also, look at what I'm thinking. I'll be there to help you be the best math problem solver you can be. Have a great summer!


P.S.  It's best to do 3 lessons every week of your summer. Pick 3 days, like Monday, Wednesday, and Friday, to practice these skills. That way you won't feel foggy in the fall when school starts. It'll be fun!

# Lesson #1

### Strategy: Make an Organized List

When we read a math question that asks us to *list all the possible correct answers* for a problem, making an **organized list** is very helpful.

What is an **organized list**, and why does it help us solve problems? An **organized list** of possible answers uses an order that makes sense to you so that you do not miss any ideas or write the same answer more than once. Let's look at an example.

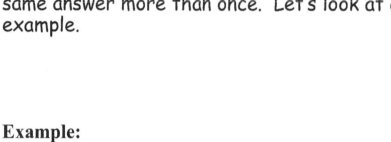

### Example:

List all of the 3-digit numbers you can make using 9, 3, and 6.

**Use an organized list** in this kind of problem. Keep one number in its place and move the other two.

I'll keep the 9 in hundreds place: 936 and 963.
I'll keep the 3 in hundreds place: 396 and 369.
I'll keep the 6 in hundreds place: 693 and 639.

### Final answer:

I have 6 different ways: 936, 963, 396, 369, 693, and 639.

Now use an **organized list** to do the next four problems.

1. List all the ways you can arrange the letters D, A, and R. It is okay if they aren't real words.

2. Mom asks you to put a quarter, a dime, and a nickel in the parking meter. The meter only takes one coin at a time. List all the ways you can "feed the meter." How many ways did you find?

3. Taylor's mom served hot dogs at an afternoon cook-out. The hot dog condiments (toppings) were ketchup, mustard, and relish. Every child put 1, 2, or 3 condiments on her hot dog. Write down all the possible ways hot dog toppings could have been used.

4. In her suitcase, Patti has a green T-shirt, a red tank top, and a pink polo shirt. She also brought navy blue shorts, denim shorts, and white shorts. List the different ways Patti can wear one top and one pair of shorts.

Use these boxes to show your work.

1.

2.

3.

4.

# Lesson #2

Use these boxes to show your work.

1. List all of the 3-digit numbers you can make using 7, 5, and 1. How many different numbers did you find?

2. A tall, skinny clown at the county fair bellowed, "There's a winner every 5 minutes!" How many winners will there be in an hour?

3. To **double an amount** is to add two of the same amount. For example, 10 doubled is 10 + 10 or 20; to double 22, write or think 22 + 22 = 44. What is 44 doubled?

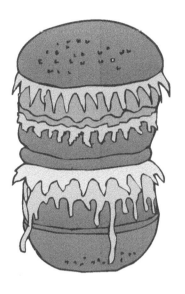

4. Draw a 7 × 4 array. Then write a multiplication problem to go with your picture.

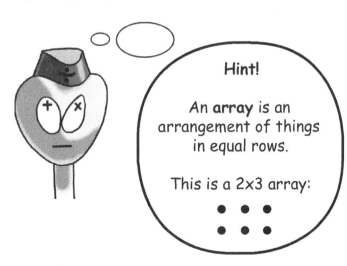

Hint!

An **array** is an arrangement of things in equal rows.

This is a 2x3 array:

# Lesson #3

## Strategy: Guess and Check

Sometimes to solve a math problem, you have to think like a detective (or a spy). Detectives follow clues that lead them to the right answer. For the **guess and check** strategy, take a guess and see if it fits all the clues by checking each one. If it does, you have solved the problem, and if it doesn't, keep trying until it works out. One way to know you have the best answer is when your answer fits every clue. Let's look at an example.

**Example:**

I am thinking of two numbers. If you add them, the sum is 17, if you subtract them, the difference is 1. What are the two numbers?

The problem tells us:

- to find 2 numbers.
- the 2 numbers will add up to 17.
- when I subtract the numbers, the answer will be 1.

First, think of a pair of numbers that equal 17 when added together. Then, subtract the smallest from the largest, in each pair, until you find the two that have a difference of 1.

Like this:

11 + 6 = 17 → 11 - 6 = 5     5 ≠ 1     This sign, ≠, means
16 + 1 = 17 → 16 - 1 = 15   15 ≠ 1    does not equal.
10 + 7 = 17 → 10 - 7 = 3     3 ≠ 1
9 + 8 = 17 → 9 - 8 = 1

**Final answer:**

The two numbers are 9 and 8 because 9 + 8 = 17 AND 9 - 8 = 1. Both clues have been checked and found to be TRUE.

**Summer Solutions© Problem Solving**                      Level 3

Now try using **guess and check** to do the next four problems.

Use these boxes to show your work.

1. **The sum of two numbers is 12 and their product is 35.** What are the numbers?

   *The **product** is the answer to a multiplication problem.*

   1.

2. **TeeJay slept a total of 16 hours Friday and Saturday. He got two more hours of sleep Saturday night than he did Friday night.** How many hours of sleep did TeeJay get each night?

   2.

3. **Mr. Cash has $6.48. He has 2 bills and 6 coins.** Show in pictures or words the bills and coins Mr. Cash has.

   3.

4. **The coins in Jacob's pocket were worth $1.04. He had 10 coins.** How many of each coin did Jacob have? (Use pennies, nickels, dimes, and quarters.)

   4.

# Lesson #4

Use these boxes to show your work.

1. **Did You Know?** Male giraffes grow up to 18 feet tall. Female giraffes grow up to 16 feet tall. Use these numbers to find the difference in the height of male and female giraffes.

**Hint!** A **difference** is the answer to a subtraction problem.

1.

2. You know that **3 feet = 1 yard**.

   Use this and the information in item 1 to **calculate** (figure out) how tall a male giraffe would be in **yards**.

2.

3. Sammi has the same number of dimes, nickels, and pennies. She has 64¢. How many of each coin does she have? Use words or pictures to show your work.

3.

4. **Dividing a number in half is easy.** Study this: What is half of 54?

- 54 = 50 + 4.
- Half of 50 is 25, and half of 4 is 2.
- So, half of 54 is 25 + 2, or 27.

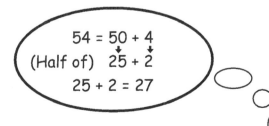

You try it. Find half of 68.

4.

# Lesson #5

**Strategy: Look for a Pattern**

Often, if we **look for a pattern** we will see a way to solve a math problem. Sometimes math problems ask us to *continue a pattern by writing what comes next*. A **pattern** is some idea that repeats. In order to write what comes next in the pattern you will first need to study the given information.

As you study it, see if there is an idea that repeats. Let's try some.

**Example:**

What number is likely to be next if the pattern continues?

15, 25, 35, 45, ____

This is a number pattern, so decide if the numbers are getting larger or smaller.

Circle one:     getting larger     getting smaller

In math, whole numbers increase (get larger) by adding or multiplying.

What would you add to 15 to get 25?      Add 10, right?
What would you add to 25 to get 35?      Add 10 again, right?
What would you add to 35 to get 45?      Add 10 again, right?

You have found a pattern!!

So, to get the number that is next in the pattern, what should you do?

If you are thinking 45 + 10 = 55, then you understand how this works.

**Final answer:**     15, 25, 35, 45, <u>55</u>

Summer Solutions© Problem Solving                                                    Level 3

Now look for a **pattern** as you solve the
next four problems.

Use these boxes to show your work.

1. Dad gives a teaspoonful of cough medicine to the new baby at 8 a.m., 12 noon, and 4 p.m. If this pattern continues, at what time should Dad give the baby cough medicine again?

   1.

2. Carrie was stringing beads to make a bracelet. She has these colors strung together: red, white, blue, white, red, white, blue, white, red. If Carrie keeps going on with this pattern, what colors will the next two beads be?

   2.

3. Decide what comes next in the pattern. Draw the next shape in the box.

   3.

4. Write the next four numbers in the pattern.

   21, 28, 35, ____, ____, ____, ____

   Use words or pictures to describe the pattern.

   4.

11

# Lesson #6

Use these boxes to show your work.

1. Nick has $1.60. Nick has 9 coins. What are they?

2. Write the next three numbers in the pattern. Read below for a little help.

60, 57, 54, _____, _____, _____

Here's one way to think about it. This is a number pattern, so decide if the numbers are getting larger or smaller.

**Circle one:**   getting larger     getting smaller

Increase means to get larger. Decrease means to get smaller.

In the box to your right, describe the pattern.

3. You know that **12 inches = 1 foot**.

   How many inches are in a 6 foot jump rope?

4. I'm thinking of a number.
   - It is a multiple of 3 AND a multiple of 5.
   - This number is less than 20 and greater than 0.

   What is the number?

   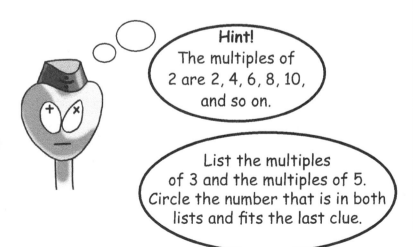

   **Hint!** The multiples of 2 are 2, 4, 6, 8, 10, and so on.

   List the multiples of 3 and the multiples of 5. Circle the number that is in both lists and fits the last clue.

## Lesson #7

**Strategy: Draw a Picture**

You don't have to be a famous artist to **draw a picture** that helps you to solve a math problem. When you **draw a picture** it helps you see the ideas you are trying to understand. The picture makes it easier for you to understand the words. Let's **draw a picture** to help us solve this problem.

**Example:**

Poppa Pete's Pizza forgot to cut the pepperoni pizzas. (Say that three times fast for fun!) So Paul cut each of the two round pizzas in half. Then, he cut each half into 3 equal pieces. After Paul cut the pizzas, how many slices could be served?

First, decide what shapes to draw. Look for hints in the problem.

"Round" is a hint to draw a circle and "two" tells how many.

Paul drew 2 circles. Now what?

Always go back to the words of the problem when you get stuck. Paul cut the pizzas in **half**. To cut something in half means **to cut it into 2 equal pieces**.

Then, Paul cut each half into 3 equal pieces.

Finally, count the total number of slices that you have drawn in your picture.

**Final answer:** 12 slices could be served.

Solve the next four problems by **drawing a picture**.

Use these boxes to show your work.

1. Kyle sliced his birthday cake into 4 rows. Each row had 5 equal slices. How many slices could be served from Kyle's cake?

2. Teisha and her friends made a long line out of 3 jump ropes. Each jump rope was 8 feet long. What was the length of the jump ropes after the children joined them?

3. Kira wanted $\frac{1}{4}$ of a jumbo chocolate chip cookie rather than $\frac{1}{3}$. She thought four is larger than three, so $\frac{1}{4}$ must be larger than $\frac{1}{3}$.

   Draw a picture to show Kira why she's really getting a smaller piece. Shade each picture and label its fraction amount. Don't forget the chocolate chips!

4. What fraction of the letters in "OHIO" are O's?

   **For example:**
   1 out of 4 letters is an "I"; that fraction is $\frac{1}{4}$.

# Lesson #8

Use these boxes to show your work.

1. I'm thinking of two numbers. The sum of the numbers is 20 and the product is 99. What are the two numbers?

1.

2. Mrs. Ross is in charge of ordering T-shirts for "It Is Rocket Science" Camp. The T-shirts can be ordered in either orange or blue. This summer you can choose either a v-neck or round neckline. Mrs. Ross decided to buy only 2 sizes, medium and large. Make an organized list of all the different types of T-shirts that need to be ordered for Science Camp.

2.

3. **Claire broke 12 oatmeal raisin cookies in half to share.** How many halves would that be? Draw a picture and write your answer in a sentence.

4. **Leo is making a pattern.** If his pattern continues, what are the next two shapes likely to be? Draw them in the box.

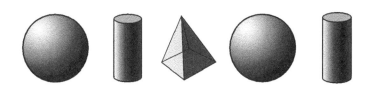

# Lesson #9

## Strategy: Work Backward

If you ever used the rewind button on a remote control, then you have a good idea about what it means to **work backward**. Using this strategy comes in handy when you know the end and the steps along the way, but you don't know how it began. If you start at the end and do the steps in reverse order, you will end up at the beginning. If you had a hard time following that, let's see how you can **work backward** to solve this problem.

### Example:

Gina came home from Mimi's Mini-Mart with $2.50. She spent $1.50 on a small popcorn and $6.00 on a Super Sub sandwich. How much money did Gina have before she went to Mimi's Mini-Mart?

Here's the simple story in reverse order:

Gina now ($2.50); buy popcorn ($1.50); buy sandwich ($6.00); go to the store with money ($ ?? ).

Here's the simple story as it happened:

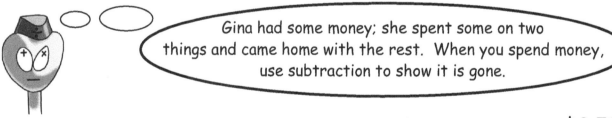

Gina had some money; she spent some on two things and came home with the rest. When you spend money, use subtraction to show it is gone.

The reverse of subtraction is addition. So if we work backward (or in reverse) and add the amounts that Gina spent to what she had at the end, we should find out what she had before she went to Mimi's Mini-Mart. Let's do that.

$2.50
$1.50
+ $6.00
_____
$10.00

### Final answer:

Gina had $10.00 before she went to Mimi's Mini-Mart.

**Work backward** to solve the next four problems.

Use these boxes to show your work.

1. Swim lessons start at 9:00 a.m. on Monday. It takes Corey 5 minutes to shower and put on some clothes. He needs 15 minutes to eat breakfast. His walk to the pool takes 10 minutes. What is the latest time that Corey can start to get ready?

   1.

2. Ayana and her little sisters love to play in the sandbox at the park. The square sandbox has a **perimeter** (the length around a shape) of 36 feet. What is the length of one side of the sandbox?

   2.

3. The perimeter of a triangle is 30 centimeters. It is an equilateral triangle, so each side is the same length. What is the measure of one side of this shape?

   3.

4. How much change should Nia get from a $10 bill when she spends $3.50 on a Veg-E Burger?

   4.

## Lesson #10

Use these boxes to show your work.

1. What fraction of the letters in "PENNSYLVANIA" are N's? Explain your answer using words and/or pictures. Do you know an equivalent fraction(same value, different name) for your answer? Write that also.

2. Guess what talent the Spitting Cobra has? Yep, you guessed it. It spits. The cobra can shoot its venom (poison) more than 6 feet (1.8 meters). Circle the measurement below that is more than 6 feet. Use your Help Pages.

      70 inches      2 meters      2 yards

Summer Solutions© Problem Solving — Level 3

3. Mrs. Smith bought one BBQ chicken dinner and two BBQ rib dinners at the Rib Burn-Off on Sunday. The BBQ chicken dinner was $8. The two rib dinners were the same price. If all three dinners cost Mrs. Smith $28, how much did each BBQ rib dinner cost?

**Hint!** Subtract the cost of a chicken dinner from the total Mrs. Smith paid. Then divide that amount by 2 (rib dinners).

4. I'm thinking of two numbers. Their sum is 16 and their product is 64. What are the two numbers?

# Lesson #11

**Strategy: Solve a Simpler Problem**

You may have noticed that every new school year there are more challenging ways to think about numbers. The numbers in your math books are also getting larger. When you read a math problem with ideas that seem too big to understand, try to **solve a simpler problem**. You probably know what to do, but it just looks too hard to figure out. Let's make this next problem simpler to solve.

**Example:**

Tee-Jay and Dee-Dee picked up aluminum cans during Riley Road's Recycle Event yesterday. Tee-Jay had 128 cans in his blue bag, but Dee-Dee was too tired to count hers. Mrs. Jones at the Drop-Off Center told them they did a great job together, bringing in a total of 252 cans to recycle. How many cans did Dee-Dee collect in her blue bag?

What do you know?
- The number of cans Tee-Jay picked up (128).
- The total number of cans both kids picked up (252).

What are you trying to find out?
- The number of cans Dee-Dee picked up.

Let's solve a similar problem with simpler numbers.

If ___ plus ___ = total, then the total minus ___ = ___.

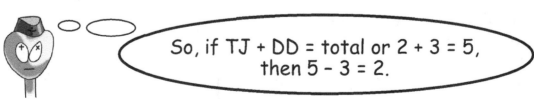

Using **simpler numbers** helped us to know that we should <u>subtract</u> to find out the amount of cans Dee-Dee collected.

Summer Solutions© Problem Solving                                    Level 3

Let's return to our example and write it this way:          252 - 128 = ?

**Final answer:**

Dee-Dee collected 124 cans in her blue bag.

See if **solving a simpler problem** helps you.

Use these boxes to show your work.

1. **Did You Know? June 1st is Children's Day in China. Children get into many events free on that day. While Ji waited in a long winding line for a free movie, he counted 182 adults.** If there were 500 people waiting for the movie, how many were children?

1.

2. **The slide at King Park is 9 feet long.** If Tommy went down the slide 25 times on Monday afternoon, how many total feet did he slide that day?

2.

**Dinosaur Den**
Adults         $49.99
Children       $19.99

3. **An estimate tells about how much.** Estimate the total cost of a visit to the Dinosaur Den for the Ellis family: Mom, Dad, and 2 children.

   $49.99 is almost   $_____.

   $19.99 is almost   $_____.

3.

4. Now find the actual cost of a visit to the Dinosaur Den Park for the Ellis family. Show your work in the box.

4.

23

# Lesson #12

Use these boxes to show your work.

1. **Did You Know?** There are 1900 active volcanoes on Earth. Within the borders of Indonesia are 75 active volcanoes. How many active volcanoes are NOT in Indonesia?

2. I'm thinking of two numbers. The product of the numbers is 42 and the difference is 1. What are the two numbers? Show your work in the box on the right.

3. What fraction of the letters in "CALIFORNIA" are A's? Explain your answer using words and/or pictures. Do you know an equivalent fraction (same value, different name) for your answer? Write that also.

3.

4. Ms. Wilson made three cherry pies for the Fourth of July. She cut each pie into 8 equal pieces. How many pieces is that? If 2 ½ pies were eaten for the holiday, how much is left?

Show your work in the box.

Remember to answer both questions.

4.

# Lesson #13

## Strategy: Make a Table

Before you ask for wood, a hammer, and nails, let's make sure you know the kind of table we mean. This kind of table is really a large T-shape that you draw on paper. Tables have rows and columns. Labels are helpful, too. Writing your ideas in this type of table (or chart) can help you organize the information in a problem so you can find an answer more easily. Sometimes it will make a pattern show up that you did not see before. Let's **make a table** to help you solve this next problem.

**Example:**

Risa was getting ready for a Swim-a-Thon at Roberto Clemente Rec Center. Starting on Monday she swam 3 laps. Tuesday she swam 2 more laps than on Monday. Each day she added 2 laps to her training from the day before. If she did this Monday through Friday, how many laps did she swim all 5 days to train for her Swim-a-Thon?

How can you put this data (information) into a T-shaped table?

The problem has days of the week and the number of laps for each.

Make a large T and write the days of the week (Monday through Friday) in the left-hand column. Then, reread the words above to place the correct number of laps swum by Risa each day in the right-hand column.

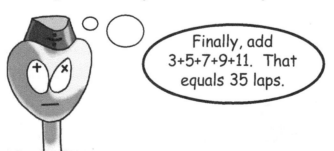

Finally, add 3+5+7+9+11. That equals 35 laps.

| Days of the Week | Laps |
|---|---|
| Monday | 3 |
| Tuesday | 5 |
| Wednesday | 7 |
| Thursday | 9 |
| Friday | 11 |

**Final answer:**
Risa swam 35 laps during the 5 days of training.

Summer Solutions© Problem Solving                                    Level 3

**Make a table** to solve the next four problems.

Use these boxes to show your work.

1. Every cabin at summer camp gets the pool for ½ hour each morning. The pool opens at 8:30 a.m. Each cabin reports on time in number order beginning with Cabin #1. If you are in Cabin #5 what time do you and your cabin mates get to swim?

1.

| Cabin # | Time |
|---------|------|
|         |      |

2. The Ohio River is 1,310 miles long. The Mississippi River is 2,340 miles long and the Tennessee River is 886 miles long. Make a table listing the names and lengths of the rivers from longest to shortest.

2.

| River | Miles |
|-------|-------|
|       |       |

3. How much longer than the Tennessee River is the Ohio River? Use the data in item 2.

3.

4. **Make a table** to solve this problem. Week 1 Malik earns $1.25. Every week this summer Malik's dad will increase his allowance by 50 cents if Malik remembers to practice his violin without being reminded. Malik doesn't have to be reminded for a whole month. What allowance does he earn in Week 4?

4.

| Week # | Amount ($) |
|--------|------------|
|        |            |
|        |            |
|        |            |
|        |            |

# Lesson #14

1. The first Friday in August is the 7th and the second Friday is the 14th. School starts the fourth Friday of August. What date will that be? Show the pattern.

Use these boxes to show your work.

1.

2. Kylie split her yard-long strawberry licorice into 3 equal pieces. Her two friends did the same with their licorice. How many smaller licorice pieces did the friends have altogether?

2.

3. Karl and Karla played lawn darts on Monday, Wednesday, and Friday. They kept adding to their score all week. On Wednesday they scored 22 points. On Friday they scored 34 points. At the end of the week, they had 101 total points. How many points must they have earned on Monday?

3.

4. A spinner has 8 equal sections. Three sections are blue, one section is yellow, and 4 sections are green. Which color is the spinner most likely to land on? Explain your answer. You could make your own spinner with paper, markers, a ruler and a paper clip.

4.

# Lesson #15

### Strategy: Write a Number Sentence

When you write a sentence you need a subject and a predicate. A **number sentence** is made up of numbers and math symbols (+ - ÷ × = < >). To use this strategy, you will turn the words of a problem into numbers and symbols. It's a snap. Take a look at the example.

### Example:

Kwame wanted to play relay races with 14 of his cousins at the family reunion. They decided to have 5 kids on a team. How many teams played relay races?

Begin to turn the words of the problem into numbers and symbols, like this:

"Kwame wanted to play relay races with 14 of his cousins." ➡ 1 + 14 = 15
"They decided to have 5 kids on a team."  ➡ 15 ÷ 5 = 3

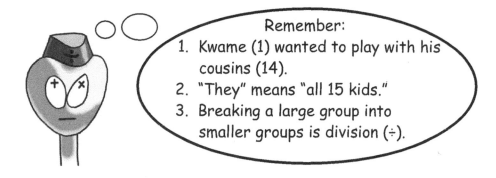

Remember:
1. Kwame (1) wanted to play with his cousins (14).
2. "They" means "all 15 kids."
3. Breaking a large group into smaller groups is division (÷).

### Final answer:

3 teams of cousins played relay races at Kwame's family reunion.

Now write a **number sentence** for the next four problems.

Use these boxes to show your work.

1. An octopus has 8 arms around its head. If there are three octopi in the tank at the Atlanta Aquarium, how many octopus arms would that be?

1.

2. **Did You Know?** Inside the body of an octopus there are 3 hearts.

   The Mills Marina has ten octopi. How many octopus hearts are pumping at Mills Marina?

3. **Mystery Number** Decide what numbers are needed and then find the **sum**.

   - Add up the number of arms on an octopus,
   - the number of eggs in 2 dozen,
   - the number of ounces in a pound,
   - and the number of feet in a yard.

   What's the sum?

4. Neriah spent $1.59 for buttered popcorn and $2.99 for chocolate-covered raisins at the Wednesday afternoon movies. Did she get any change back from her $5 bill? If so, how much?

# Lesson #16

1. Jared and his mom got on at the first floor elevator at St. John's Hospital. The elevator went up 10 floors, then down 5 floors. Then they got out. At what floor did they get off the elevator?

   Hint! If you get on the 1st floor elevator and go up two floors, you are on the 3rd floor.

2. Lashawna won 55 tickets playing games at the Game-O-Rama. She spent 20 tickets on prizes. Then she won 35 more tickets. She spent 30 tickets on another prize and gave the rest to her little brother. How many tickets did her little brother receive?

Use these boxes to show your work.

1.

2.

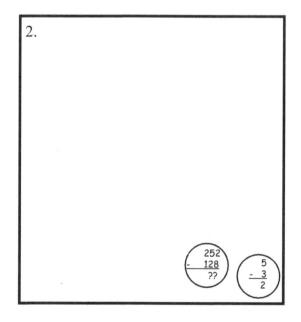

3. What number is twenty-four more than 29 AND eighteen less than 71?

Hint! Write two different number sentences to find the one right answer.

4. You know that the vowels are A, E, I, O, U, and sometimes Y. The other letters of the alphabet are consonants. Let's say you have one set of magnetic alphabet letters. You throw them in a brown paper bag. You place your hand in the bag and pull out one letter. What's the probability it will be a vowel? Explain your answer in words and/or pictures.

Hint! The probability of rolling a 3 on one number cube is 1 out of 6 or 1/6.

# Lesson #17

### Strategy: Use Logical Reasoning

**Logical reasoning** is basically common sense. **Logical** means "sensible." **Reasoning** is "a way of thinking." **Logical reasoning** is done one step at a time until you see the whole answer. Problem solving in this way is like solving a jigsaw puzzle. Take a look at the example.

### Example:

Izzy is 2 years older than Bizzy. Bizzy is twice as old as Lizzy. If Lizzy is 6 years old, how old are Bizzy and Izzy? **Don't get dizzy!**

Doing this one step at a time will help us discover their ages.

Reread the problem to find out what you know about their ages.

    Lizzy is 6 years old.
    Bizzy is twice Lizzy's age. Twice means *2 times*.
    Izzy is 2 years older than Bizzy. Two years older means *plus 2*.

Start with what you know and put the pieces together.

    Bizzy is two times Lizzy's age.
    Izzy is Bizzy's age plus two.

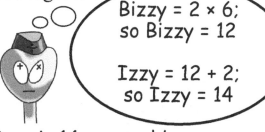

Bizzy = 2 × 6;
so Bizzy = 12

Izzy = 12 + 2;
so Izzy = 14

**Final answer:**

    Bizzy is 12 years old and Izzy is 14 years old.

Use **logical reasoning** as you solve the next four problems.

Use these boxes to show your work.

1. At Stellar Stadium, hot dogs are 75¢ more than plain cheese pizza. Cheeseburgers are $1.25 more than hot dogs. A large bag of chips is half the price of a cheeseburger. Find the cost of each food if the price of pizza is $1.

1.

2. Pam put a yellow, a white, a pink and a blue daisy in a vase. The daisies had 16, 20, 24, and 30 petals each.

- The pink daisy had 10 more petals than the blue.
- The yellow daisy had 6 fewer petals than the pink daisy.
- The white daisy had the fewest petals.

How many petals did each color daisy have?

**Hint!** Order the numbers from fewest to most petals. Then find the differences between the numbers.

3. **Rip Van Winkle fell asleep for twenty years.** Is that about 8,000 or 800 days? Use estimation to help you solve this problem.

4. **Dan is half as old as Nan. Jan is 2 years younger than Dan.** If Jan is 7 years old, how old are Nan and Dan? *Come on now, get a plan!*

# Lesson #18

Use these boxes to show your work.

1. **Did You Know?** The usual school day in Australia goes from 9 a.m. to 3:30 p.m. Australian students eat lunch at school. In Brazil, students attend school from 7 a.m. to noon. Students go home to share lunch, the most important meal of the day, with their families. In which country do the students spend a longer time at school? How much longer?

2. **Mystery Number**
   Decide what numbers are needed and then find the **sum**.

   - Add up the number of eggs in a half dozen,
   - the number of days in a week,
   - the number of inches in 3 feet,
   - and the number of cents in a quarter.

   What's the sum?

Summer Solutions© Problem Solving — Level 3

3. Devon has been collecting rain water this summer in a clear, plastic cup. On Sunday, he uses his metric ruler and learns that he has a total of 9 centimeters of water in the cup. For the next week there is no rain and the temperature is between 88° and 92°F. On Monday, the cup held $8\frac{1}{2}$ cm of water. If $\frac{1}{2}$ cm of water evaporated every day after that, how much water would be in the cup Friday evening?

3.

| Day | Rain Water (cm) |
| --- | --- |
|  |  |

4. Get ready for the Water Balloon Toss! The Blue Team used 16 cups of water to fill up water balloons. The Red Team used 9 pints of water to fill up balloons. The Green Team filled up balloons with 4 quarts of water. Which team used more than a gallon of water? Explain your answer in the box with words or pictures.

2 cups = 1 pint,
2 pints = 1 quart, and
4 quarts = 1 gallon.

4.

# Lesson #19

## Combining Strategies

To get from one side of the pool to the other, you swim a little and dog-paddle the rest of the way. This is called **combining strategies**. You can do this with math problems, too. From time to time, you will see a strategy suggestion for a problem. However, always know that there are many ways to think about and solve a problem. For the rest of the lessons, use the strategies alone or together to help find the answers.

Use these boxes to show your work.

1. At breakfast on July 5th, the temperature was 92°F. Around noon it dropped 5° as colder winds moved across the lake. By dinner time, the sun was back out and the temperature had risen 3°. What was the temperature at dinner time?

**Hint!**
Rising temperature → use addition;
Dropping temperature → use subtraction

2. Study the rectangular prism and the pyramid. Use a table to compare the number of faces and edges on each.

| Solid Shape | # of faces | # of edges |
|---|---|---|
| pyramid | | |
| rectangular prism | | |

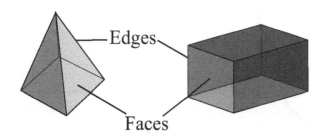
Edges
Faces

3. **Greater than, Less than, or Equal to?** Decide if these amounts are greater than (>) each other, less than (<) each other, or equal to (=) each other and draw the right sign in the box.

one-third of 15      one-fourth of 12

First, draw 15 small shapes and put them into 3 equal groups (because you are trying to find 1/3 of that number). How many are in one group? _____

Draw your shapes here: 

Then, draw 12 small shapes and put them into 4 equal groups (because you are trying to find ¼ of that number). How many are in one group? _____

Draw your shapes here:

Finally, use >, <, or = to compare the two numbers. Write a number sentence to show your final answer.

one-third of 15      one-fourth of 12

(Write >, <, or =.)

4. Draw a 9 x 3 array. Then write a multiplication problem to go with your picture.

4.

# Lesson #20

1. **Greater than, Less than, or Equal to?** Decide if these amounts are greater than (>) each other, less than (<) each other or equal to (=) each other and draw the right sign in the box.

    one half of 16　☐　one eighth of 16

    Use these boxes to show your work.

    | one-half of 16 | one-eighth of 16 |
    |---|---|
    |  |  |

2. The sum of what number and thirty-three is eighty? Write a **number sentence** and use a question mark or little square in place of the words "what number." Do a **simpler problem** to help you know what to do with the number words.

    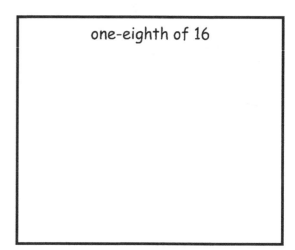

    **Hint!** The sum of what number and 2 is 5?

    ☐ + 2 = 5

3. A van picks up Alyia for summer church camp on Mondays, Wednesdays, and Fridays. It is a 15 minute ride to camp in the morning and a 20 minute ride home in the afternoon. How much time does Alyia spend each week traveling to and from camp? Give your answer in hours and minutes. Show your work.

4. Helen, Simon, Josie, and Gracie love to write poems. Over the summer the children wrote 8, 12, 14, and 20 poems each. Josie wrote as many as Simon and Gracie combined. Gracie wrote 6 fewer than Helen. How many poems did each of them write? Show your thinking in words, numbers, or pictures.

# Lesson #21

## Sort Angles - Acute, Right, and Obtuse

Today you are going to classify (sort) clocks. Before we do that, let's review the three types of angles. **Right angles** measure exactly 90° and look like an "L". **Acute angles** measure less than 90° (∠). **Obtuse angles** measure more than 90° (⌒). You may be wondering what angles have to do with clocks. Well, it was timely of you to ask. Clocks with a "face" and "hands" are called analog clocks. You don't have to remember that. But, you probably noticed that the minute hand and hour hand meet in the center of the clock and form an angle. So every minute of the day, the hands of the clock are coming together to make different angles. Now here's your job.

Use these boxes to show your work.

1.
- Head each column with the name of one of the three types of angles.
- Decide what the time is on each clock.
- Then, decide which type of angle the hands of the clock make.
- Finally, write the clock time in the column of your chart that goes with the correct angle. (**Example:** When the hands show 3 o'clock or 3:00 they are forming a right angle. Write 3:00 under Right Angle.)

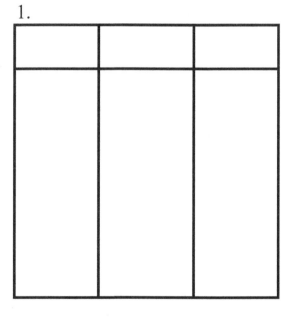

2. The Brown family, Jones family, and Fields family each are taking road trips this summer. Their trips will be 312 miles, 255 miles, and 284 miles. The Fields family will travel 28 fewer miles than the Jones family. The Fields family will travel 29 miles farther than the Browns. How many miles will each family travel this summer?

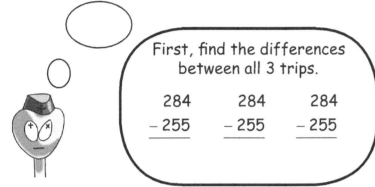

First, find the differences between all 3 trips.

```
  284      284      284
- 255    - 255    - 255
```

2.

3. Remember that
   **1 pound equals 16 ounces** (1 lb = 16 oz).

   Which weighs less, a bag of frozen corn at 33 ounces or a box of frozen peas at 1 pound 15 ounces? How much less? Explain.

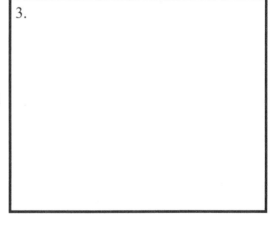

3.

4. What is half of 108?

4.

Summer Solutions© Problem Solving                    Level 3

# Lesson #22

## Use a Graph/Make a Graph

Remember **pictographs** and **bar graphs** from this past school year? These types of graphs help us see lots of data all at once. Study the pictograph and bar graph below. Answer the questions that follow. Then you will be ready to make these types of graphs on your own. There will be help along the way.

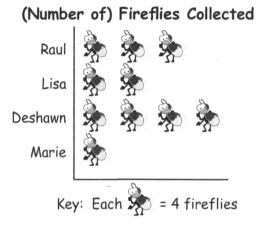

1. How many fireflies did the four children collect in all? _____

   How many more fireflies did Deshawn collect than Lisa? _____

   Circle the picture that would equal 2 fireflies. 🪰🪰 / 🪰 / 🪰

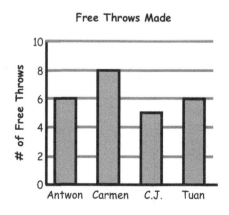

2. Which players got the same amount of free throws? _____

   What is the **range** (highest # minus lowest #) of the data? _____

   What is the **mode** (the # that occurs most often) of this data? _____

3. Make a pictograph using the data in the table.
   - Label the months.
   - Draw the correct number of symbols.

   | Month | # of sunny days |
   |-------|-----------------|
   | June | 12 |
   | July | 15 |
   | August | 24 |

   Have each ☼ stand for 6 days of sunshine.

   How many days is ☼ ? _____

4. Make a bar graph using the data in the tally chart. Remember to title your graph.

   **Beach Treasures Found**

   | Beach Glass | ||||  ||||  ||||  ||||  |||| |
   |-------------|---------------------------|
   | Snail Shells | ||||  ||| |
   | Red Rocks | ||||  ||||  |||| |
   | Flat Stones | ||||  ||||  ||||  |||| |

   How many treasures were found at the beach in all?

   _____

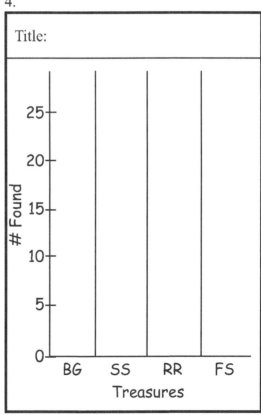

## Lesson #23

Use these boxes to show your work.

1. What fraction of the letters in "ILLINOIS" are consonants? Use words and/or pictures. Do you know an equivalent fraction (same value, different name) for your answer? Write that also.

2. What are the next two numbers in the following pattern likely to be?

$\frac{1}{2}$, 1, $1\frac{1}{2}$, 2, $2\frac{1}{2}$, 3, _____, _____

3. A tubful of water balloons were needed for the Taylor Family Reunion. Mr. Taylor bought 9 jumbo bags of balloons. Each jumbo bag contained 55 balloons. If only 10 balloons broke, how many water balloons were able to be used for the family reunion?

9 groups of 55, then minus 10.

4. Use the T-shaped table. The heading for the left column is 12. The heading for the right column is 15. Solve these six problems. The letter stands for the missing number. Write the letter in the column that matches your answer.

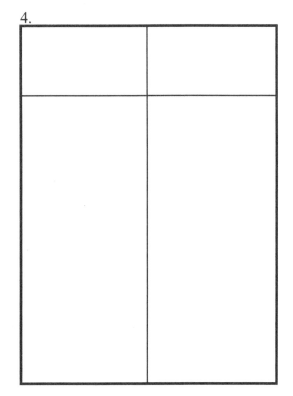

$2 + R = 14$; R is 12, so I'd write R in the left column.

$4 \times M = 48$      $B \times 2 = 30$
$(2 \times 10) - 5 = Z$      $42 - 27 = H$
$48 - K = 36$      $2 \times 3 \times 2 = G$

**Lesson #24**

## Notice Unimportant Information

Not every number or word is needed to find the answer to a math problem. Let's re-visit item 3 in Lesson #18. Read it over again and notice what information is interesting, but not really necessary.

Devon has been collecting rain water this summer in a clear, plastic cup. On Sunday, he uses his metric ruler and learns that he has a total of 9 centimeters of water in the cup. For the next week there is no rain and the temperature is between 88° and 92°F. On Monday, the cup held 8 ½ cm of water. If ½ cm of water evaporated every day after that, how much water would be in the cup Friday evening?

1. With a marker or crayon, cross out the unnecessary information in the above paragraph.

Use these boxes to show your work.

2. A hexagon has a perimeter of 60 inches. Every side of the hexagon is the same length. What is the length of one of the sides?

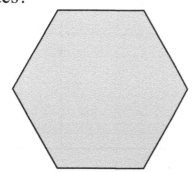

2.

3. In August the sun rises about 1 minute later every day. The morning sun rises at 5:30 on August 1st, 5:31 on August 2nd, and 5:32 on August 3rd. If this pattern continues, what time will the sun rise on August 6th? You can use a table to help show your pattern.

3.

| Date | Time |
|---|---|
|  |  |

4. Mr. Mixed-up has three shapes: a triangle, a square, and an octagon. He can't remember what color he is supposed to paint each one. Here are some clues to help him. **The red shape has twice as many sides as the blue one. The blue shape has one more side than the purple one.** Give the name of the shape, the number of sides, and its color.

4.

| Shape | # of sides | Color |
|---|---|---|
|  |  |  |

## Lesson #25

1. Draw a picture to show 19 ÷ 4. Remember to write a left over amount as a remainder.

   Use these boxes to show your work.

   1.

2. Did you know the Three Little Pigs have been learning about taking care of our planet? Now, they want to make houses out of recycled materials. They can choose to make the walls from glass, cans, or old tires. The roof can be made from plastic or Styrofoam. Their house needs walls and a roof. List the different combinations that can be designed. (for each combination, choose only 1 wall and 1 material for the roof.)

   2.

3. I'm thinking of two numbers. The sum of the numbers is 14 and the product is 48. What are the two numbers?

3.

4. The downtown bus runs every 20 minutes beginning at 9:00 a.m. Here is the beginning of the bus schedule: 9:00, 9:20, 9:40, 10:00. What is the time of the first bus after 12 noon?

4.

# Lesson #26

1. **Here is a T-shaped table.** Solve these six problems. **The letters stand for 36 or 24.** Write the letter in the column that matches your answer.

   3 × 4 × 3 = Y     2 × 6 × 2 = S
   3 dozen = Q      (8 × 5) − 4 = A
   6 + (2 × 9) = F   2 dozen = T

   Use these boxes to show your work.

   1.
   | 36 | 24 |
   |----|----|
   |    |    |

2. **Perry Township has 4 softball teams in the Independence Day Parade this year. Each team has 9 players.** How many softball players are in the parade?

3. Every giant step that Nina takes is equal to 3 feet. Simon Says Nina may take 11 giant steps. How many feet is that?

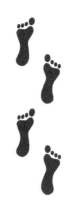

**Act this one out** by playing Simon Says!

4. Auntie Cassie gave her nieces some choices. They could visit the Children's Museum or go to the roller skating rink. They could do this in the morning or in the afternoon. How many choices do Auntie Cassie's nieces have? List them all.

# Lesson #27

Use these boxes to show your work.

1. Jing-Wei is making a tower of blocks. She is following a pattern as she builds. From the bottom up the blocks are pink, blue, blue, purple, pink, blue, blue, purple, and so on. What color will the 12th block be?

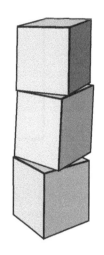

2. Aunty Kymie built an octagon-shaped picnic table at her woodworking class. She had to make sure each side was exactly the same length. The perimeter of the picnic table was 24 feet. What was the length of one side of the table?

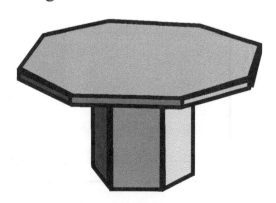

3. Mason thinks 12 fourths are equal to 3 wholes. Is he correct?

4. The summer tee-ball playoffs ran for two days in August. A full house at Thornton Park would be 1,200 fans. On August 15th the park was 89 tickets short of a full house. On August 16th the park was 224 tickets less than a full house. How many fans attended the playoff games during the two days?

# Lesson #28

1. Uncle Vincie is trying to eat healthier meals and is losing weight. He weighed 205 pounds at the end of May. If he wants to lose 5 pounds each month, how much will he weigh by the end of August? Use a table to find the answer.

2. Leah was allowed one suitcase weighing 40 pounds on her flight to New York City. She spent 30 minutes taking things out of her luggage until the weight was okay. She removed a computer weighing 8 lbs. and some books weighing 4 lbs. She also had to take out 2 lbs. of sandals. What did the suitcase weigh before Leah started taking things out of the bag?

Use these boxes to show your work.

1.

| Month | Weight |
|-------|--------|
|       |        |

2.

3. On the first day of summer camp, Allie has to pick a bunk mate from her cabin and then decide if she wants the top or bottom bunk bed. Her choices of bunk mates are Morgan, Mia, and Madison. Show all the different sleeping arrangements from which Allie can choose.

3.

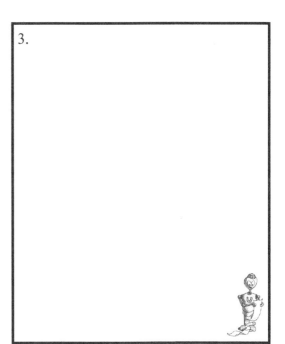

4. Draw a picture to show $29 \div 6$. Remember to write a left over amount as a remainder.

4.

# Lesson #29

Use these boxes to show your work.

1. A chess board has an area of 64 squares. The game board **is a square.** Without checking an actual board, decide how many squares must be along every side of the board. Use words and/or pictures.

2. Here's the game that Bill, Gill, and Jill played. Run as fast as possible up to the yellow line, then jump. Whoever jumps the farthest is the winner. Cousin Will measured these:

   - 1 yard
   - 2 feet 10 inches
   - 38 inches

   Gill's jump was the shortest. Bill jumped 2 inches farther than Jill. How far did each person jump?

   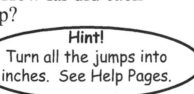
   **Hint!** Turn all the jumps into inches. See Help Pages.

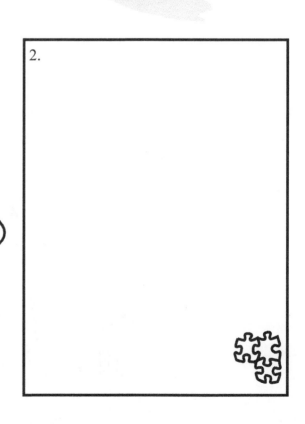

3. Mr. Grant has $11.27. **He has 3 bills and 7 coins.** Show in pictures or words the bills and coins Mr. Grant has.

4. Lenny came home from the movies with $4.75. He spent $2.50 on a large root beer, $4 on French fries, and $4.25 on a corn dog. How much money did Lenny have before he went to the movies?

# Lesson #30

Use these boxes to show your work.

1. Papa Russ planted 3 rows of sweet corn in April this year. If there were 25 corn stalks in each row, how many stalks would that be in all?

1.

2. Lincoln has 10 coins. The number of dimes equals the number of nickels. He has more pennies than dimes. He has 49 cents. How many of each coin does Lincoln have?

2.

3. A spinner has 10 equal sections. Five sections are red, two sections are pink, one section is orange and two sections are violet. Which color is the spinner most likely to land on? It is equally likely that the spinner will land on which two colors? Explain your answer.

There are 2 questions to answer here.

3.

4. Erika's grandma bought 60 inches of blue ribbon to wrap party favors. She cut the ribbon in half. Then she cut each half into three equal pieces. How many pieces did she have in the end? How long was one piece?

4.

# Level 3

# Problem Solving

## Help Pages

# Help Pages

## Vocabulary

### General Terms

**Denominator** — the bottom number of a fraction
Example: $\frac{1}{4}$; The denominator is 4.

**Difference** — the result or answer to a subtraction problem
Example: The difference of 5 and 1 is 4.

**Fraction** — a part of a whole   Example: This box has 4 parts; 1 part is shaded. $\frac{1}{4}$

**Numerator** — the top number of a fraction   Example: $\frac{1}{4}$; The numerator is 1.

**Product** — the result or answer to a multiplication problem
Example: The product of 5 and 3 is 15.

**Quotient** — the result or answer to a division problem
Example: The quotient of 8 and 2 is 4.

**Remainder** — the part left over when one number can't be divided exactly by another

**Sum** — the result or answer to an addition problem   Example: The sum of 5 and 2 is 7.

**Symbols** — signs         = → equals              ≠ → does not equal
                            > → is greater than    < → is less than

### Geometry

**Acute Angle** — an angle measuring less than 90°

**Area** — the size of a surface; area is always given in square units (feet², meters²,...)

**Congruent** — figures with the same shape and the same size

**Diameter** — the widest distance across a circle; the diameter always passes through the center

**Edge** — where 2 faces meet on a 3-dimensional (solid) shape

**Equilateral** — all sides have the same measurement

**Face** — the flat part of a 3-dimensional (solid) shape

**Line of Symmetry** — a line along which a figure can be folded so that the two halves match exactly

**Obtuse Angle** — an angle measuring more than 90°

**Perimeter** — the distance around the outside of a polygon

**Radius** — the distance from any point on a circle to the center; the radius is half of the diameter

**Right Angle** — an angle measuring exactly 90°

**Similar** — figures having the same shape, but different sizes

# Help Pages

Vocabulary

| Geometry — Polygons ||||
|---|---|---|---|
| Number of Sides | Name | Number of Sides | Name |
| 3 △ | Triangle | 6 ⬡ | Hexagon |
| 4 ▢ | Quadrilateral | 8 ⯃ | Octagon |
| 5 ⬠ | Pentagon | | |

| Geometry — Solid Figures ||
|---|---|
| Cylinder — | Rectangular Prism — |
| Pyramid — | Sphere — |

| Measurement — Relationships ||
|---|---|
| **Volume** | **Distance** |
| 3 teaspoons in a tablespoon | 36 inches in a yard |
| 2 cups in a pint | 1760 yards in a mile |
| 2 pints in a quart | 5280 feet in a mile |
| 4 quarts in a gallon | 100 centimeters in a meter |
| **Weight** | 1000 millimeters in a meter |
| 16 ounces in a pound | **Temperature** |
| 2000 pounds in a ton | 0° Celsius – Freezing Point |
| **Time** | 100° Celsius – Boiling Point |
| 10 years in a decade | 32° Fahrenheit – Freezing Point |
| 100 years in a century | 212° Fahrenheit – Boiling Point |

**Problem Solving Terms**

**Arrange** — put in a certain order

**Array** — an arrangement of things in rows or columns

**Calculate** — figure out

**Classify** — sort

**Data** — information

# Help Pages

Vocabulary

| Problem Solving Terms (continued) |
|---|
| **Decrease** — get smaller |
| **Double** — add 2 of the same amount; 10 doubled = 10 + 10 = 20 |
| **Equivalent** — same value or amount, different name |
| **Estimate** — a close guess |
| **Height** — how high something is |
| **Increase** — get larger |
| **Length** — how long something is |
| **Multiples** — the multiples of 3 are 3, 6, 9, 12, etc. |
| **Pattern** — an idea that repeats |
| **Probablility** — the chance of someting happening |
| **Strategy** — plan |
| **Table** — a chart with rows and columns |
| **Width** — how wide something is |
| **Statistics** |
| **Mode** — the number that occurs most often in a group of numbers. The mode is found by counting how many times each number occurs in the list. The number that occurs more than any other is the mode. Some groups of numbers have more than one mode. |
| Example: The mode of 77, ⑨③, 85, ⑨③, 77, 81, ⑨③, and 71 is **93**. (93 is the mode because it occurs more than the others.) |
| **Range** — in a set of data, the highest number minus the lowest number. |
| Example: The range of 77, 93, 85, 93, 77, 81, 93, and 71 is **22**. (93 − 71 = 22) |

Place Value

| Whole Numbers |
|---|
| 2 7 1, 4 0 5 <br> Hundred Thousands / Ten Thousands / Thousands / Hundreds / Tens / Ones |
| The number above is read: two hundred seventy-one thousand, four hundred five. |

Summer Solutions© Problem Solving                    Level 3

# Help Pages

Problem Solving

## Strategies

### Make an Organized List

An **organized list** of possible answers for a problem uses an order that makes sense to you so that you do not miss any ideas or write the same answer more than once.

### Guess and Check

For the **guess and check** strategy, take a guess and see if it fits all the clues by checking each one. If it does, you have solved the problem. If it doesn't, keep trying until it works out. One way to know you have the best answer is when your answer fits <u>every</u> clue.

### Look for a Pattern

Sometimes math problems ask us to *continue a pattern by writing what comes next*. A **pattern** is an idea that repeats. In order to write what comes next in the pattern, you will first need to study the given information. As you study it, see if there is an idea that repeats.

### Draw a Picture

When you **draw a picture** it helps you see the ideas you are trying to understand. The picture makes it easier to understand the words.

### Work Backward

Using this strategy comes in handy when you know the end of a problem and the steps along the way, but you don't know how the problem began. If you start at the end and do the steps in reverse order you will end up at the beginning.

### Solve a Simpler Problem

When you read a math problem with ideas that seem too big to understand, try to **solve a simpler problem**. Instead of giving up or skipping that problem, replace the harder numbers with easier ones.

### Make a Table

**Tables** have columns and rows. Labels are helpful too. Writing your ideas in this type of table (or chart) can help you organize the information in a problem so you can find an answer more easily. Sometimes it will make a pattern show up that you did not see before.

### Write a Number Sentence

A **number sentence** is made up of numbers and math symbols (+ − × ÷ > < =). To use this strategy you will turn the words of a problem into numbers and symbols.

Summer Solutions© Problem Solving                                                    Level 3

Problem Solving        Help Pages

### Strategies (continued)

**Use Logical Reasoning** - **Logical reasoning** is basically common sense. **Logical** means "sensible." **Reasoning** is "a way of thinking." **Logical reasoning** is done one step at a time until you see the whole answer.

Solved Examples

### Whole Numbers (continued)

When we **round numbers**, we are estimating them. This means we focus on a particular place value, and decide if that digit is closer to the next highest number (round up) or to the next lower number (keep the same). It might be helpful to look at the place-value chart on page 285.

Example: Round 347 to the tens place.

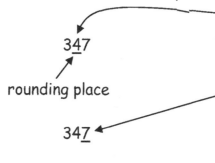

Since 7 is greater than 5, the rounding place is <u>increased by 1</u>.

**350**

1. Identify the place that you want to round to. What number is in that place? (4)
2. Look at the digit to its right. (7)
3. If this digit is 5 or greater, increase the number in the rounding place by 1. (round up) If the digit is less than 5, keep the number in the rounding place the same.
4. Replace all digits to the right of the rounding place with zeroes.

Here is another example of rounding whole numbers.

Example: Round 4,826 to the hundreds place.

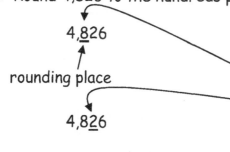

Since 2 is less than 5, the rounding place <u>stays the same</u>.

**4,800**

1. Identify the place that you want to round to. What number is in that place? (8)
2. Look at the digit to its right.
3. If this digit is 5 or greater, increase the number in the rounding place by 1. (round up) If the digit is less than 5, keep the number in the rounding place the same.
4. Replace all digits to the right of the rounding place with zeroes.

Summer Solutions© Problem Solving                                    Level 3

# Help Pages

Solved Examples

## Whole Numbers (continued)

**When adding or subtracting whole numbers,** first the numbers must be lined-up on the right. Starting with the ones place, add (or subtract) the numbers; when adding, if the answer has 2 digits, write the ones digit and regroup the tens digit (for subtraction, it may also be necessary to regroup first). Then, add (or subtract) the numbers in the tens place. Continue with the hundreds, etc.

Look at these examples of **addition**.

Examples:  Find the sum of 314 and 12.                    Add 6,478 and 1,843.

```
  314
+  12
-----
  326
```

1. Line up the numbers on the right.
2. Beginning with the ones place, add. Regroup if necessary.
3. Repeat with the tens place.
4. Continue this process with the hundreds place, etc.

```
  1 1 1
  6,478
 +1,843
 ------
  8,321
```

Use the following examples of **subtraction** to help you.

Examples:  Subtract 37 from 93.

```
  8 13
  9 3
- 3 7
-----
  5 6
```

1. Begin with the ones place. Check to see if you need to regroup. Since 7 is larger than 3, you must regroup to 8 tens and 13 ones.
2. Now look at the tens place. Since 3 is less than 8, you do not need to regroup.
3. Subtract each place value beginning with the ones.

Find the difference of 425 and 233.

```
  3 12
  4 2 5
- 2 3 3
-------
  1 9 2
```

1. Begin with the ones place. Check to see if you need to regroup. Since 3 is less than 5, you do not need to regroup.
2. Now look at the tens place. Check to see if you need to regroup. Since 3 is larger than 2, you must regroup to 3 hundreds and 12 tens.
3. Now look at the hundreds place. Since 2 is less than 3, you are ready to subtract.
4. Subtract each place value beginning with the ones.

# Help Pages

## Solved Examples

### Whole Numbers (continued)

Sometimes when doing subtraction, you must **subtract from zero**. This always requires regrouping. Use the examples below to help you.

Examples:  Subtract 261 from 500.

$$\begin{array}{r} \overset{\phantom{0}\phantom{0}9}{\underset{\phantom{0}}{4\ \cancel{10}\ 10}} \\ \cancel{5}\ \cancel{0}\ \cancel{0} \\ -\ 2\ 6\ 1 \\ \hline 2\ 3\ 9 \end{array}$$

1. Begin with the ones place. Since 1 is less than 0, you must regroup. You must continue to the hundreds place, and then begin regrouping.
2. Regroup the hundreds place to 4 hundreds and 10 tens.
3. Then, regroup the tens place to 9 tens and 10 ones.
4. Finally, subtract each place value beginning with the ones.

Find the difference between 600 and 238.

$$\begin{array}{r} \overset{\phantom{0}\phantom{0}9}{\underset{\phantom{0}}{5\ \cancel{10}\ 10}} \\ \cancel{6}\ \cancel{0}\ \cancel{0} \\ -\ 2\ 3\ 8 \\ \hline 3\ 6\ 2 \end{array}$$

**Multiplication** is a quicker way to add groups of numbers. The sign (×) for multiplication is read "times." The answer to a multiplication problem is called the product. Use the examples below to help you understand multiplication.

Examples:  3 × 5 is read "three times five."

It means *3 groups of 5*  or  5 + 5 + 5.

3 × 5 = 5 + 5 + 5 = 15

The product of 3 × 5 is **15**.

4 × 7 is read "four times seven."

It means *4 groups of 7*  or  7 + 7 + 7 + 7.

4 × 7 = 7 + 7 + 7 + 7 = 28

The product of 4 × 7 is **28**.

## Help Pages

Solved Examples

### Whole Numbers (continued)

It is very important that you memorize your **multiplication facts**. This table will help you, but only until you memorize them!

To use this table, choose a number in the top gray box and multiply it by a number in the left gray box. Follow both with your finger (down and across) until they meet. The number in that box is the product.

An example is shown for you: 2 × 3 = 6

| × | 0 | 1 | 2 | 3 | 4 | 5 | 6 | 7 | 8 | 9 | 10 |
|---|---|---|---|---|---|---|---|---|---|---|----|
| 0 | 0 | 0 | 0 | 0 | 0 | 0 | 0 | 0 | 0 | 0 | 0 |
| 1 | 0 | 1 | 2 | 3 | 4 | 5 | 6 | 7 | 8 | 9 | 10 |
| 2 | 0 | 2 | 4 | 6 | 8 | 10 | 12 | 14 | 16 | 18 | 20 |
| 3 | 0 | 3 | 6 | 9 | 12 | 15 | 18 | 21 | 24 | 27 | 30 |
| 4 | 0 | 4 | 8 | 12 | 16 | 20 | 24 | 28 | 32 | 36 | 40 |
| 5 | 0 | 5 | 10 | 15 | 20 | 25 | 30 | 35 | 40 | 45 | 50 |
| 6 | 0 | 6 | 12 | 18 | 24 | 30 | 36 | 42 | 48 | 54 | 60 |
| 7 | 0 | 7 | 14 | 21 | 28 | 35 | 42 | 49 | 56 | 63 | 70 |
| 8 | 0 | 8 | 16 | 24 | 32 | 40 | 48 | 56 | 64 | 72 | 80 |
| 9 | 0 | 9 | 18 | 27 | 36 | 45 | 54 | 63 | 72 | 81 | 90 |
| 10 | 0 | 10 | 20 | 30 | 40 | 50 | 60 | 70 | 80 | 90 | 100 |

Summer Solutions© Problem Solving    Level 3

# Help Pages

## Solved Examples

### Whole Numbers (continued)

When **multiplying multi-digit whole numbers**, it is important to know your multiplication facts.  Follow the steps and the examples below.

Examples:  Multiply 23 by 5.

$\overset{1}{2}3$     $3 \times 5 = 15$ ones or 1 ten and 5 ones
$\times 5$
$\overline{115}$    $2 \times 5 = 10$ tens + 1 ten (regrouped)
          or 11 tens.

1. Line up the numbers on the right.
2. Multiply the digits in the ones place.  Regroup if necessary.
3. Multiply the digits in the tens place.  Add any regrouped tens.
4. Repeat step 3 for the hundreds place, etc.

Find the product of 314 and 3.

$3\overset{1}{1}4$     $4 \times 3 = 12$ ones or 1 ten and 2 ones.
$\times\ 3$
$\overline{942}$    $1 \times 3 = 3$ tens + 1 ten (regrouped) or 4 tens.
          $3 \times 3 = 9$ hundreds.

**Division** is the opposite of multiplication.  The symbols for division are ÷ and $\overline{)}$ and are read "divided by."  The answer to a division problem is called the quotient.  Remember that multiplication is a way of adding groups to get their total.  Think of division as the reverse of this.  In a division problem you already know the total and the number in each group.  You want to know how many groups there are.  Follow the examples below.

Examples:  Find the quotient of 12 ÷ 3.    (12 items divided into groups of 3)

   The total number is 12.    △ △ △ △ △ △ △ △ △ △ △ △
   Each group contains 3.    (△ △ △)(△ △ △)(△ △ △)(△ △ △)
   How many groups are there?   There are 4 groups.
                    12 ÷ 3 = **4**

   Divide 10 by 2.    (10 items divided into groups of 2)
   The total number is 10.    △ △ △ △ △ △ △ △ △ △
   Each group contains 2.    (△ △)(△ △)(△ △)(△ △)(△ △)
   How many groups are there?   There are 5 groups.
                    10 ÷ 2 = **5**

# Help Pages

## Solved Examples

### Whole Numbers (continued)

Sometimes when you are dividing, there are items left over that do not make a whole group. These left-over items are called the **remainder**. When this happens, we say that "the whole cannot be divided evenly by that number."

Example: What is 16 divided by 5?   (16 items divided into groups of 5)

The total number is 16.

Each group contains 5.

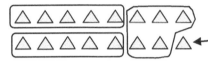

How many groups are there?   There are 3 groups, but there is 1 left over. The remainder is 1.

16 ÷ 5 = **3 R1**   (This is read "3 remainder 1.")

---

The next group of examples involves **long division using one-digit divisors with remainders**. You already know how to divide single-digit numbers. This process helps you to be able to divide numbers with multiple digits.

Example: Divide 37 by 4.

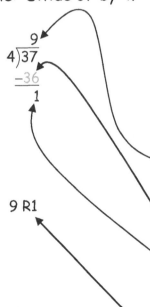

1. In this problem, 37 is the dividend and 4 is the divisor. You're going to look at each digit in the dividend, starting on the left.
2. Ask yourself if the divisor (4) goes into the left-most digit in the dividend (3). It doesn't, so keep going to the right.
3. Does the divisor (4) go into the two left-most digits (37)? It does. How many times does 4 go into 37? (9 times)
4. Multiply 4 × 9 (product = 36).
5. Subtract 36 from 37 (difference = 1). There's nothing left to bring down from above. Once this number is smaller than the divisor, it is called the remainder and the problem is finished. The remainder is 1.
6. Write the answer (above the top line) with the remainder. (9 R1)

# Help Pages

Solved Examples

## Whole Numbers (continued)

Example: What is 556 divided by 6?

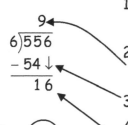

92 R4

1. Ask yourself if the divisor (6) goes into the left-most digit in the dividend (5). It doesn't, so keep going to the right.
2. Does the divisor (6) go into the two left-most digits (55)? It does. How many times does 6 go into 55? (9 times)
3. Multiply 6 × 9 (product is 54).
4. Subtract 54 from 55. (1) Bring down the 6 ones from the first line. This leaves 16 left from the original 556.
5. Ask yourself if the divisor (6) goes into 16. It does. How many times does 6 go into 16? (2)
6. Multiply 6 × 2 (product is 12).
7. Subtract 12 from 16 (result is 4). There's nothing left to bring down from above. Once this number is smaller than the divisor, it is called the remainder and the problem is finished. The remainder is 4.
8. Write the answer with the remainder. (92 R 4)

Remember: The remainder can NEVER be larger than the divisor!

## Fractions

A **fraction** is used to represent part of a whole. The top number in a fraction is called the **numerator** and represents the part. The bottom number in a fraction is called the **denominator** and represents the whole.

 The whole rectangle has 6 sections.
Only 1 section is shaded.
This can be shown as the fraction $\frac{1}{6}$.

$\frac{1}{6}$ shaded part (numerator) / total parts (denominator)

To **add (or subtract) fractions with the same denominator**, simply add (or subtract) the numerators, keeping the same denominator.

Examples: $\frac{3}{5} + \frac{1}{5} = \frac{4}{5}$   $\frac{8}{9} - \frac{1}{9} = \frac{7}{9}$

# Help Pages

Solved Examples

## Decimals

**Adding and subtracting decimals** is very similar to adding or subtracting whole numbers. The main difference is that you have to line-up the decimal points in the numbers before you begin. Add zeros if necessary, so that all of the numbers have the same number of digits after the decimal point. Before you subtract, remember to check to see if you must regroup. When you're finished adding (or subtracting), bring the decimal straight down into your answer.

Example: Find the sum of 4.25 and 2.31.

```
  4.25
+ 2.31
  ----
  6.56
```

1. Line up the decimal points. Add zeroes as needed.
2. Add (or subtract) the decimals.
3. Add (or subtract) the whole numbers.
4. Bring the decimal point straight down.

Example: Subtract 4.8 from 7.4.

```
  6 14
  7.4
- 4.8
  ---
  2.6
```

## Geometry

The **perimeter** of a polygon is the distance around the outside of the figure. To find the perimeter, add the lengths of the sides of the figure. Be sure to label your answer.

Perimeter = sum of the sides

Example: Find the perimeter of the rectangle below.

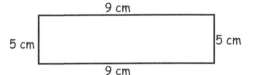

Perimeter = 5 cm + 9 cm + 5 cm + 9 cm
Perimeter = 28 cm

Example: Find the perimeter of the regular pentagon below.

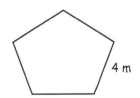

A pentagon has 5 sides. Each of the sides is 4 m long.

P = 4 m + 4 m + 4 m + 4 m + 4 m
P = 5 × 4 m
P = 20 m

# Help Pages

## Who Knows???

Sides in a Quadrilateral? ................................. (4)

Sides in a Pentagon? ..................................... (5)

Sides in a Hexagon? ...................................... (6)

Sides in an Octagon? ..................................... (8)

Inches in a foot? ........................................... (12)

Feet in a yard? .............................................. (3)

Inches in a yard? ........................................... (36)

Ounces in a pound? ....................................... (16)

Pounds in a ton? ........................................... (2000)

Cups in a pint? ............................................. (2)

Pints in a quart? ........................................... (2)

Quarts in a gallon? ........................................ (4)

Years in a decade? ........................................ (10)

Figures with the same size and shape? ................ (congruent)

Figures with same shape, but different size? ........ (similar)

Answer to an addition problem? ..................... (sum)

Answer to a subtraction problem? .... (difference)

Answer to a multiplication problem? ...... (product)

Answer to a division problem? ............... (quotient)

# Level 3

# Problem Solving

Answers to Lessons

## Lesson #1

| | | | |
|---|---|---|---|
| 1 | DAR  ADR  RAD<br>DRA  ARD  RDA | 3 | K = ketchup   R = relish<br>M = mustard<br>Possible ways to use toppings:<br>K only, M only, R only,<br>KM, KR, MR, KMR<br>There are 7 ways. |
| 2 | Q = quarter<br>D = dime<br>N = nickel<br><br>QND   NQD   DQN<br>QDN   NDQ   DNQ<br>There are six ways. | 4 | green T + navy shorts<br>green T + denim shorts<br>green T + white shorts<br>red tank + navy shorts<br>red tank + denim shorts<br>red tank + white shorts<br>pink polo + navy shorts<br>pink polo + denim shorts<br>pink polo + white shorts |

## Lesson #2

| | | | |
|---|---|---|---|
| 1 | 751   517   175<br>715   571   157<br>There are 6 different numbers. | 3 | $\phantom{+\,}44$<br>$\underline{+\,44}$<br>$\phantom{+\,}88$ |
| 2 | 1 hour = 60 minutes<br>There are twelve 5's in 60, so there will be 12 winners. | 4 | (7 rows × 4 columns of stars)<br>7 x 4 = 28    4 x 7 = 28 |

## Lesson #3

| | | | |
|---|---|---|---|
| 1 | The numbers are 7 and 5.<br>7 + 5 = 12 (sum)<br>7 x 5 = 35 (product) | 3 | 2 bills and 6 coins = $6.48<br>$5 dollar bill<br>$1 dollar bill<br>one quarter, two dimes, and three pennies |
| 2 | <u>Hours of Sleep</u>:<br>Friday         7 hours<br>Saturday     9 hours<br>9 is 2 more than 7.<br>9 + 7 = 16 total hours | 4 | 10 coins worth $1.04<br>three quarters    $0.75<br>two dimes           $0.20<br>one nickel           $0.05<br>4 pennies            <u>$0.04</u><br>                              $1.04 |

## Lesson #4

| | | | |
|---|---|---|---|
| 1 |   18 feet (male)<br>− 16 feet (female)<br>   2 feet difference | 3 | 4 dimes = $0.40<br>4 nickels = $0.20<br>4 pennies = <u>$0.04</u><br>           $0.64 |
| 2 | 3 feet = 1 yard<br>male giraffe is 18 feet tall<br>18 ÷ 3 = 6 yards<br>The male giraffe must be 6 yards tall. | 4 | Half of 68:<br>Half of 60 is 30.<br>Half of 8 is 4.<br>30 + 4 is 34.<br>Half of 68 is 34. |

|   | Lesson #5 |   |   |
|---|---|---|---|
| 1 | 8:00am + 4 = 12 noon<br>12 noon + 4 = 4:00 pm<br>4:00pm + 4 = 8:00 pm<br><br>The pattern is "every 4 hours." Dad should give the medicine at 8:00 pm. | 3 | Square is the next shape in the pattern.<br><br>(The pattern is 3 sides, 4 sides, 5 sides, 3 sides, 4 sides, and so on. |
| 2 | The pattern is<br>red, white, blue, white,<br>red, white, blue, white,<br>red, ____, ____<br>White and blue are the next two colors. | 4 | 21,28,35, 42,49,56,63<br><br>The pattern is "add 7."<br>(Or, each number is 7 more than the last one.) |

|   | Lesson #6 |   |   |
|---|---|---|---|
| 1 | 5 quarters  $1.25<br>3 dimes  $0.30<br>1 nickel  $0.05<br>9 coins  $1.60<br>     Or<br>1 fifty-cent piece  $0.50<br>3 quarters  $0.75<br>2 dimes  $0.20<br>3 nickels  $0.15<br>9 coins  $1.60 | 3 | 12 inches = 1 foot<br><br>6 feet     12<br>           x 6<br>           72 inches<br><br>72 inches are in a 6 ft. jump rope. |
| 2 | The numbers are getting smaller by 3.<br>60, 57, 54, 51, 48, 45<br>(Or, subtract 3 to get the next number.) | 4 | Multiples of 3:<br>3, 6, 9, 12, **15**, 18, 21, . . .<br>Multiples of 5:<br>5, 10, **15**, 20, 25, 30, . . .<br>15 is greater than zero and less than 20. |

| | | | Lesson #7 | |
|---|---|---|---|---|
| 1 | 4 rows of 5 equal slices. 20 slices of cake can be served. | 3 | $\frac{1}{4} < \frac{1}{3}$ | |
| 2 | 8 + 8 + 8 = 24 feet of jump ropes | 4 | <u>O</u>H I <u>O</u><br><br>2 of the 4 letters are O's, so the fraction is $\frac{2}{4}$ or $\frac{1}{2}$. | |

| | | | Lesson #8 | |
|---|---|---|---|---|
| 1 | 11 + 9 = 20 (sum)<br>11 × 9 = 99 (product)<br><br>The numbers are 11 and 9. | 3 | 12 cookies cut in half equal 24 halves. | |
| 2 | O = orange   B = blue<br>V = v-neck   R = round neck<br>M = medium   L = large<br><br>OVM     BVM<br>OVL     BVL<br>ORM    BRM<br>ORL     BRL<br><br>8 different types of shirts | 4 | sphere, cylinder, pyramid, sphere, cylinder, _____, _____ | |

## Lesson #9

| | | | |
|---|---|---|---|
| 1 | 9:00 am<br>−    5 minutes<br>8:55 am<br>−   15 minutes<br>8:40 am<br>−   10 minutes<br>8:30 am is the latest. | 3 | perimeter = 30 cm<br><br>30 ÷ 3 = 10 cm<br>One side of the shape measures 10 cm. |
| 2 | perimeter = 36 ft.<br><br>36 ÷ 4 = 9 ft.<br>The length of one side of the sandbox is 9 feet. | 4 | $10.00<br>− $3.50<br>  $6.50<br><br>Nia should get $6.50 in change. |

## Lesson #10

| | | | |
|---|---|---|---|
| 1 | PE**NN**SYLVA**N**IA<br><br>3 of 12 letters are N's, so the fraction is $\frac{3}{12}$ or $\frac{1}{4}$. | 3 | $28.00<br>−   8.00<br>$20.00 ÷ 2 = ?<br><br>Half of 20 = 10, so the rib dinners cost $10 each. |
| 2 | (2 meters)<br><br>2 is more than 1.8<br><br>2 yards = 6 feet<br>6 feet = 72 inches | 4 | 8 + 8 = 16 (sum)<br>8 × 8 = 64 (product)<br><br>The two numbers are 8 and 8. |

## Lesson #11

| | | | |
|---|---|---|---|
| 1 | 500<br>- 182<br>318 children were waiting in line for the movie. | 3 | $49.99 is almost $50.00<br>$19.99 is almost $20.00<br><br>Mom + Dad: 50 + 50 = 100<br>2 children: 20 + 20 = 40<br>100 + 40 = 140.00<br>The family will spend about $140.00. |
| 2 | 25<br>x 9<br>225<br><br>Tommy slid 225 feet. | 4 | $49.99<br>49.99<br>19.99<br>+ 19.99<br>$139.96 actual cost |

## Lesson #12

| | | | |
|---|---|---|---|
| 1 | 1900<br>- 75<br>1825 active volcanoes are not located in Indonesia. | 3 | C$\underline{A}$LIFORNI$\underline{A}$<br>2 of 10 or $\frac{2}{10}$<br><br>$\frac{2}{10} = \frac{1}{5}$ |
| 2 | 7 x 6 = 42 (product)<br>7 - 6 = 1 (difference)<br><br>The numbers are 7 and 6. | 4 | 3 pies are cut into eighths; the shaded part was eaten.<br><br>1) There are 24 pieces of pie.<br>2) If $2\frac{1}{2}$ pies were eaten,<br>$\frac{4}{8}$ or $\frac{1}{2}$ of a pie is left. |

|   | Lesson #13 |   |   |
|---|---|---|---|
| 1 | Cabin #   Time<br>1   8:30 am<br>2   9:00 am<br>3   9:30 am<br>4   10:00 am<br>5   10:30 am<br>Cabin #5 gets to swim at 10:30 am. | 3 | 1310<br>-  886<br>  424<br><br>The Ohio River is 424 miles longer than the Tennessee River. |
| 2 | River       Miles<br>Mississippi   2,340<br>Ohio          1,310<br>Tennessee      886 | 4 | Week #       Amount ($)<br>1                 1.25<br>2   (1.25 + .50) 1.75<br>3   (1.75 + .50) 2.25<br>4   (2.75 + .50) 2.75<br><br>Malik will earn $2.75 in week #4. |

|   | Lesson #14 |   |   |
|---|---|---|---|
| 1 | Aug. 7, Aug. 14, ___, ___,<br><br>The 4th Friday is Aug. 28th. The pattern is "add 7." | 3 | 101           34<br>- 34         + 22<br>  67   or    56<br>- 22          101<br>  45         - 56<br>45 points     45<br>were earned on Monday. |
| 2 | Kylie      __ __ __<br>Friend 1   __ __ __<br>Friend 2   __ __ __<br><br>There will be 9 smaller licorice pieces in all. | 4 | GGGG, BBB, Y<br>Green has the most spaces, so the spinner is most likely to land on green. |

| | | Lesson #15 | | |
|---|---|---|---|---|
| 1 | | 8 × 3 = 24<br>There would be 24 octopus arms. | 3 | Add the numbers:<br>  8 (octopus arms)<br>24 (2 dozen eggs)<br>16 (ounces in a pound)<br>+ 3 (feet in a yard)<br>51 is the sum. |
| 2 | | 10 × 3 = 30<br>There are 30 octopus hearts pumping at the Mills Marina. | 4 | $5.00<br>− 2.99<br>$ 2.01<br>− 1.59<br>$0.42 Neriah's change |

| | | Lesson #16 | | |
|---|---|---|---|---|
| 1 | |   1    1st floor<br>+ 10  up 10 floors<br> 11<br>− 5  down 5 floors<br>  6<br>They got off the elevator at the 6th floor. | 3 |  29    71<br>+ 24  − 18<br> 53    53<br><br>The number is 53. |
| 2 | |  55    (start)<br>− 20  (spent 20 tickets)<br> 35<br>+ 35  (won 35 tickets)<br> 70<br>− 30  (spent 30 tickets)<br> 40<br>Lashawna gave 40 tickets to her little brother. | 4 | There are 26 letters in the alphabet. The probability of pulling a vowel is 5 out of 26 (6 out of 26 if you include y as a vowel.)<br><br>Answer: $\frac{5}{26}$ or $\frac{6}{26}$ |

| Lesson #17 |||||
|---|---|---|---|---|
| 1 | pizza = $1.00<br>hot dogs = $1.75<br>    (pizza + $ .75)<br>cheeseburgers = $3.00<br>    (hotdog + $1.25)<br>chips = $1.50<br>    ($\frac{1}{2}$ cheeseburger;<br>    half of $3.00 is $1.50) | 3 ||| One year is 365 days, and 365 is close to 400.  20 groups of 400 equals 20 x 400 = 8,000<br>(800 is only 2 x 400)<br>Answer: 8,000 days |
| 2 | Daisy Petals:<br>white: 16 (fewest)<br>pink: 30 (10 + blue)<br>blue: 20<br>yellow: 24 (pink − 6) | 4 ||| Jan = 7 (given)<br>Dan = 9 (Jan's age + 2)<br>Nan = 18 (Dan's age x 2) |

| Lesson #18 |||||
|---|---|---|---|---|
| 1 | Australia:<br>9am − 3:30pm = 6 $\frac{1}{2}$ hours<br>Brazil:<br>7am − 12noon = 5 hours<br>6 $\frac{1}{2}$ − 5 = 1 $\frac{1}{2}$<br>Australian students spend 1 $\frac{1}{2}$ hours longer at school. | 3 | Day | Rainwater (centimeters) ||
| ::: | ::: | ::: | Monday | 8 $\frac{1}{2}$ ||
| ::: | ::: | ::: | Tuesday | 8 ||
| ::: | ::: | ::: | Wednesday | 7 $\frac{1}{2}$ ||
| ::: | ::: | ::: | Thursday | 7 ||
| ::: | ::: | ::: | (Friday) | 6 $\frac{1}{2}$ ||
| 2 | Add: 6 (eggs in $\frac{1}{2}$ dozen)<br>          7 (days in a week)<br>         36 (inches in 3 feet)<br>        +25 (a quarter)<br>         74 is the sum. | 4 | Team | Water | Amount |
| ::: | ::: | ::: | Blue | 16 c = | 1 gal. |
| ::: | ::: | ::: | (Red | 9 pts. > | 1 gal.) |
| ::: | ::: | ::: | Green | 4 qts. = | 1 gal. |
| ::: | ::: | ::: | 8 pts. = 4 qts. = 1 gal.<br>9 pts. > 8 pts. |||

## Lesson #19

**1.**
92° F
− 5  (drops 5°)
87° F
+ 3  (rises 3°)
90° F at dinner time

**2.**

| Shape | Faces | Edges |
|---|---|---|
| Pyramid | 5 | 8 |
| Rectangular Prism | 6 | 12 |

**3.**
$\frac{1}{3}$ of 15 = 5

$\frac{1}{4}$ of 12 = 3

$\frac{1}{3}$ of 15 > $\frac{1}{4}$ of 12

**4.**
$9 \times 3 = 27 \quad 3 \times 9 = 27$

## Lesson #20

**1.**
$\frac{1}{2}$ of 16 = 8

$\frac{1}{8}$ of 16 = 2

$\frac{1}{2}$ of 16 > $\frac{1}{8}$ of 16

**2.**
? + 33 = 80
80
− 33
47
The number is 47.

**3.**
15    morning ride
+ 20   afternoon ride
35

35 (M) + 35 (W) + 35 (F) = 105 minutes

105 min. = 60 + 45 or 1 hour and 45 minutes

**4.**
Helen (H) = 14
Simon (S) = 12
Josie (J) = S + G = 20
Gracie (G) = H − 6 = 8
If Gracie wrote 8 poems, then Simon wrote 12, and Josie wrote 20.

| Lesson #21 ||||
|---|---|---|---|
| 1 | Right Angles: 3:00, 9:00<br>Acute Angles: 1:00, 2:00<br>Obtuse Angles: 7:00, 4:00 | 3 | frozen corn: <u>33 oz.</u><br>frozen peas: 1 lb., 16 oz.<br>1 pound = 16 ounces, so<br>16 oz. + 15 oz. = <u>31 oz.</u><br>Frozen peas weigh two ounces less than corn. |
| 2 | 312 (J)   312    284 (F)<br>- 284 (F) - 255   - 255 (B)<br>    28      57       29<br>Jones: 312 miles<br>Fields: 284 miles<br>Browns: 255 miles | 4 | half of 100 = 50<br>half of 8 = 4<br><br>$\frac{1}{2}$ of 108 = 54 |

| Lesson #22 ||||
|---|---|---|---|
| 1 | They collect 40 fireflies in all. Deshawn collected 8 more than Lisa.<br><br>= 2 fireflies | 3 | Sunny Summer Days chart: June 2 suns, July 2½ suns, August 4 suns. Key: Each sun stands for 6 days.<br><br>½ sun = 3 days |
| 2 | Antwon and Tuan got the same amount of free throws.<br><br>Range = 3 (8 – 5 = 3)<br>Mode = 6 (occurs most often) | 4 | 66 treasures were found.<br><br>Title: Beach Treasures Found bar graph (BG ~24, SS ~8, RR ~15, FS ~19) |

88

| Lesson #23 |||||
|---|---|---|---|---|
| 1 | ILLINOIS<br><br>4 out of 8 or $\frac{4}{8}$ are consonants.<br><br>$\frac{4}{8} = \frac{1}{2}$ | 3 | 55 balloons in each bag<br>× 9 jumbo bags<br>495 total balloons<br>− 10 balloons that broke<br>485 were able to be used. ||
| 2 | $3\frac{1}{2}$ and 4 are the next two numbers. | 4 | 12<br>M<br>K<br>G | 15<br>B<br>Z<br>H |

| Lesson #24 |||||
|---|---|---|---|---|
| 1 | ~~On Sunday, he uses a metric ruler and learns that he has a total of 9cm of water in the cup. For the next week, there is no rain and the temperature is between 88° and 92° F.~~ | 3 | Date | Time |
|   |   |   | August 1 | 5:30 |
|   |   |   | August 2 | 5:31 |
|   |   |   | August 3 | 5:32 |
|   |   |   | August 4 | 5:33 |
|   |   |   | August 5 | 5:34 |
|   |   |   | (August 6 | 5:35) |
| 2 | 60 ÷ 6 = 10<br>One side is 10 inches long. | 4 | Shape \| Sides \| Color<br>triangle \| 3 \| purple   $3 = \frac{1}{2}$ of 6<br>square \| 4 \| blue   $4 = 1 + 3$<br>octagon \| 8 \| red ||

## Lesson #25

| | 1 | | 3 | |
|---|---|---|---|---|
| 1 | (dots: 4 columns of 5, circled vertically)<br><br>19 divided by 4 = 4 r 3 | | 3 | Numbers  Sum  Product<br>1, 48     49    48<br>2, 24     26    48<br>3, 16     19    48<br>4, 12     16    48<br>(6, 8     14    48)<br><br>The numbers are 6 and 8. |
| 2 | Walls        Roof<br>glass        plastic<br>cans         plastic<br>old tires    plastic<br>glass        Styrofoam<br>cans         Styrofoam<br>old tires    Styrofoam | | 4 | 9:00, 9:20, 9:40,<br>10:00, 10:20, 10:40,<br>11:00, 11:20, 11:40,<br>12:00, (12:20)<br><br>The first bus after 12 noon is at 12:20pm. |

## Lesson #26

| | 1 | | 3 | |
|---|---|---|---|---|
| 1 | 36    24<br>Y     F<br>Q     S<br>A     T | | 3 | 3 × 11 = 33<br><br>Nina moves 33 feet. |
| 2 | 4 × 9 = 36<br><br>There are 36 softball players. | | 4 | <u>The nieces have 4 choices</u>:<br>-Children's Museum morning;<br>-Children's Museum afternoon;<br>-Roller Skating morning;<br>-Roller Skating afternoon. |

## Lesson #27

**1.** 1 pink, 2 blue, 3 blue, 4 purple, 5 pink, 6 blue, 7 blue, 8 purple, 9 pink, 10 blue, 11 blue, <u>12 purple</u>

The 12th block will be purple.

**2.** An octagon has 8 sides.
$24 \div 8 = 3$
One side of the table is 3 feet long.

**3.** Yes, three wholes = $\frac{12}{4}$.
$12 \div 4 = 3$

**4.**
```
Aug. 15    Aug. 16    Both
 1200       1200      1111
 - 89       -224      +976
 ----       ----      ----
 1111        976      2087
```
2087 fans attended during the two days.

## Lesson #28

**1.**
| Month | Weight |
|---|---|
| May | 205 lbs. |
| June | 200 lbs. |
| July | 195 lbs. |
| <u>August</u> | <u>190 lbs.</u> |

**2.**
```
  40
 + 8   computer
  48
 + 4   books
  52
 + 2   sandals
  54
```
Leah's suitcase weighed 54 pounds in the beginning.

**3.**
<u>top bunk</u>
bottom bunk

| <u>Allie</u> | <u>Allie</u> | <u>Allie</u> |
|---|---|---|
| Morgan | Mia | Madison |

| <u>Morgan</u> | <u>Mia</u> | <u>Madison</u> |
|---|---|---|
| Allie | Allie | Allie |

**4.** $29 \div 6 = 4 \text{ r } 5$

## Lesson #29

| | | | |
|---|---|---|---|
| 1 | area = length x width<br>Both numbers have to be the same because the chess board is a square. So, since $8 \times 8 = 64$,<br>there will be 8 squares along each side. | 3 | $5 + $5 + $1 = $11<br>N + N + N + N + N + P + P = 27¢<br>3 bills and 7 coins = $11.27 |
| 2 | 1 yard = 36 inches<br>2 feet, 10 in. = 34 inches<br>38 inches = 38 inches<br>Bill = 38 in. (Jill + 2 inches)<br>Gill = 34 in. (shortest)<br>Jill = 36 in. | 4 |   4.74   came home with<br>  2.50   root beer<br>  4.00   fries<br>+ 4.25   corn dog<br>$15.50  Lenny had this |

## Lesson #30

| | | | |
|---|---|---|---|
| 1 | $3 \times 25 = 75$<br><br>75 stalks in all | 3 | The spinner will most likely land on red. It is equally likely to land on violet or pink. |
| 2 | D  D  D  N  N<br>N  P  P  P  P<br>10 + 10 + 10 = 30<br>  5 + 5 + 5 = 15<br>1 + 1 + 1 + 1 = 4<br>This adds up to 49¢. | 4 | 60 inches of ribbon "cut in half" is $60 \div 2 = 30$.<br>Grandma cut each half into three equal pieces. $30 \div 3 = 10$.<br>Two halves cut into 3 pieces each, makes 6 pieces. So, there were 6 pieces, and each was 10 inches long. |